INNOVATIVE DEVELOPMENT

Global Hawk
AND DarkStar

**Their Advanced Concept
Technology Demonstration
Program Experience**

**Executive
Summary**

Jeffrey A. Drezner
Robert S. Leonard

T0149967

Prepared for the United States Air Force

RAND
Project AIR FORCE

The research reported here was sponsored by the United States Air Force under Contract F49642-01-C-0003. Further information may be obtained from the Strategic Planning Division, Directorate of Plans, Hq USAF.

ISBN: 0-8330-3111-2

Published 2002 by RAND
1700 Main Street, P.O. Box 2138, Santa Monica, CA 90407-2138
1200 South Hayes Street, Arlington, VA 22202-5050
201 North Craig Street, Suite 102, Pittsburgh, PA 15213-1516
RAND URL: http://www.rand.org/
To order RAND documents or to obtain additional information, contact Distribution Services: Telephone: (310) 451-7002;
Fax: (310) 451-6915; Email: order@rand.org

The High-Altitude Endurance Unmanned Aerial Vehicle (HAE UAV) Advanced Concept Technology Demonstration (ACTD) program incorporated a number of innovative elements into its development strategy. As a condition of conducting this ACTD, Congress required that an independent third party study its implementation. RAND was chosen for this role and has been following the HAE UAV ACTD program since its inception.[1]

The joint program was conducted until October 1998 under the direction of the Defense Advanced Research Projects Agency (DARPA) and the Defense Airborne Reconnaissance Office (DARO) and was then conducted under the U.S. Air Force through the completion of the ACTD. RAND's initial research covering Phase I and most of Phase II was sponsored by DARPA; the Air Force sponsored RAND's effort through the completion of the ACTD.

The core objective of the research was twofold: to understand how the innovative acquisition strategy used in the HAE UAV ACTD program affected program execution and outcomes, and to draw lessons from this experience that could be applicable to the wider acquisition community. Four reports were written at the conclusion of the ACTD. Three of these documents addressed specific aspects of the

---

[1]See Geoffrey Sommer, Giles K. Smith, John L. Birkler, and James R. Chiesa, *The Global Hawk Unmanned Aerial Vehicle Acquisition Process: A Summary of Phase I Experience,* MR-809-DARPA, Santa Monica: RAND, 1997; and Jeffrey A. Drezner, Geoffrey Sommer, and Robert S. Leonard, *Innovative Management in the DARPA High Altitude Endurance Unmanned Aerial Vehicle Program: Phase II Experience,* MR-1054-DARPA, Santa Monica: RAND, 1999.

HAE UAV ACTD program: a description of the program's activity content and a comparative analysis to other development efforts; transition management–related issues; and an analysis of the flight test program.

This report, the fourth in the series, is an executive summary that covers all aspects of the program. It summarizes our major research findings; identifies which elements of the innovative acquisition strategy worked well and which worked poorly; and makes recommendations for improvement. Detailed information and support can be found in our companion documents.

This research was sponsored by the Global Hawk System Joint Program Office (GHSPO) in the Aeronautical Systems Center, Reconnaissance Air Vehicle (ASC/RAV) directorate of Air Force Materiel Command (AFMC). It was conducted within RAND's Project AIR FORCE, a federally funded research and development center sponsored by the Air Force.

Reports in this series are:

MR-1473-AF, *Innovative Development: Global Hawk and DarkStar—Their Advanced Concept Technology Demonstrator Program Experience, Executive Summary,* Jeffrey A. Drezner, Robert S. Leonard

MR-1474-AF, *Innovative Development: Global Hawk and DarkStar— HAE UAV ACTD Program Description and Comparative Analysis,* Robert S. Leonard, Jeffrey A. Drezner

MR-1475-AF, *Innovative Development: Global Hawk and DarkStar— Flight Test in the HAE UAV ACTD Program,* Jeffrey A. Drezner, Robert S. Leonard

MR-1476-AF, *Innovative Development: Global Hawk and DarkStar— Transitions Within and Out of the HAE UAV ACTD Program,* Jeffrey A. Drezner, Robert S. Leonard

PROJECT AIR FORCE

Project AIR FORCE, a division of RAND, is the Air Force federally funded research and development center (FFRDC) for studies and analyses. It provides the Air Force with independent analyses of policy alternatives affecting the development, employment, combat readiness, and support of current and future air and space forces. Research is performed in four programs: Aerospace Force Development; Manpower, Personnel, and Training; Resource Management; and Strategy and Doctrine.

CONTENTS

FIGURES

TABLES

ACKNOWLEDGMENTS

This research would not have been possible without the cooperation of officials associated with the program in the U.S. Air Force, the Office of the Secretary of Defense, the Defense Advanced Research Projects Agency, and industry. Special thanks are due to government program office personnel and to the contractors who provided information and spent considerable time discussing the HAE UAV ACTD program.

We would also like to extend our thanks to Geoffrey Sommer, ex officio project team member, who provided suggestions and observations during the research and an unofficial review of the draft reports. Our formal technical reviewers, Giles Smith and Frank Fernandez, provided excellent reviews of the draft reports.

We would also like to thank Natalie Crawford and Timothy Bonds in Project AIR FORCE for providing their time and resources to ensure that this research was completed to the highest-quality standards.

Any errors are the sole responsibility of the authors.

ACC	Air Combat Command
ACTD	Advanced Concept Technology Demonstration
CGS	Common ground segment
CY	Calendar year
D&E	Demonstration and evaluation
DARO	Defense Airborne Reconnaissance Office
DARPA	Defense Advanced Research Projects Agency
Dem/val	Demonstration/validation
DoD	Department of Defense
DT&E	Development test and evaluation
EMD	Engineering and manufacturing development
EO/IR	Electro-optical/infrared
FAR	Federal Acquisition Regulation
FSD	Full-scale development
FY	Fiscal year
GHSPO	Global Hawk System Program Office

HAE UAV	High-Altitude Endurance Unmanned Aerial Vehicle
IOT&E	Initial operational test and evaluation
IPT	Integrated Product Team
JFCOM	Joint Forces Command
LCRS	Launch, control, and recovery station
LWF	Lightweight Fighter
MAE	Medium-Altitude Endurance
MDAP	Major Defense Acquisition Program
MOU	Memorandum of understanding
MUA	Military utility assessment
OSD	Office of the Secretary of Defense
OTA	Other Transaction Authority
TY	Then year
UAV	Unmanned aerial vehicle
UFP	Unit Flyaway Price

Chapter One

INTRODUCTION AND OVERVIEW

The United States has seen a three-decade-long history of poor outcomes in unmanned aerial vehicle (UAV) development efforts. UAV and tactical surveillance/reconnaissance programs have a history of failure involving inadequate integration of sensor, platform, and ground elements, together with unit costs far exceeding what operators have been willing to pay. This history motivated the unique management approach adopted at the beginning of the High-Altitude Endurance Unmanned Aerial Vehicle (HAE UAV) program.

There has also been a long history of efforts to improve the efficiency and effectiveness of acquisition policy, processes, and management for all weapon system types. Capturing the experience from ongoing or recently completed efforts employing nonstandard or innovative acquisition strategies can facilitate such improvements. This research contributes to that effort. Our core objective was twofold: to understand how the innovative acquisition strategy used in the HAE UAV Advanced Concept Technology Demonstration (ACTD) program affected program execution and outcomes, and to draw lessons from this experience that could be applicable to the wider acquisition community.

The HAE UAV ACTD Phase I and Phase II experience has been documented in previously published reports. Four reports were written at the conclusion of the ACTD. Three of these documents addressed specific aspects of the HAE UAV ACTD program: a description of the program's activity content and comparative analysis to other development efforts; transition management–related issues; and an analysis of the flight test program. This report, the fourth in the series, is an executive summary that covers all aspects of the program. It summarizes our major research findings; identifies which elements of the innovative acquisition strategy worked well and which worked poorly; and makes recommendations for improvement.

The remainder of this chapter outlines the HAE UAV ACTD program structure and provides a brief overview of how the program evolved. Chapter Two discusses the effects of the innovative acquisition strategy on key program events and outcomes. Chapter Three compares the experience of the HAE UAV ACTD program to that of other prototype and acquisition programs. Chapter Four draws some general lessons and makes several recommendations to improve the application of the strategy.

HAE UAV ACTD PROGRAM STRUCTURE

The HAE UAV ACTD program consisted of two distinct air vehicles and their respective ground segments. The Tier II+ was a conventionally configured UAV; the Tier III– incorporated low-observable features into its design. Early in the program, the concept of a common ground segment was introduced. The program was initiated by the Defense Airborne Reconnaissance Office (DARO) and the Defense Advanced Research Projects Agency (DARPA), two defense agencies that are not normally in the business of developing new weapon systems. DARPA, an agency charged with technological innovation, was given program management responsibility at the program's inception. DARPA was expected to complete the design and build the first two examples of each system as well as to prove the basic flightworthiness of each. Both development efforts would then transfer to the Air Force. The U.S. Air Force, which initially had no stated requirement, budget, or interest in either system, was to complete the ACTD.

Tier II+ Global Hawk Program Plan

The HAE UAV Tier II+ program plan consisted of four phases, as depicted in Figure 1.1. According to the HAE UAV Phase I solicitation dated June 1, 1994, the planned program structure was as follows:

RAND*MR1473-1.1*

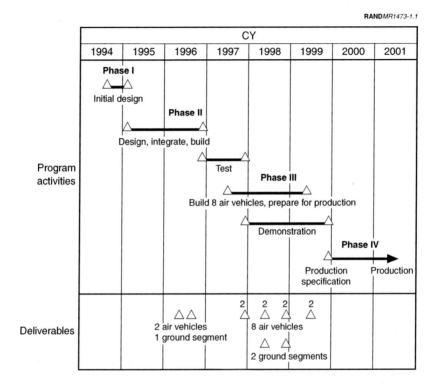

Figure 1.1—Tier II+ Schedule, Original Program

- **Phase I:** A six-month effort by three contractor teams to conduct a System Objective Review and a Preliminary System Specification Review.

- **Phase II:** A 27-month effort by two contractor teams to design and develop the Tier II+ system; complete the definition of the system specification and all interfaces; produce a prototype system; and successfully complete initial flight testing. The products were to be two prototype air vehicles, one set of sensors, a prototype ground segment, and a support segment capable of demonstrating initial system performance.

- **Phase III:** A 36-month effort by a single contractor team with the primary objective of the successful operational demonstration of the Tier II+ system. The products were to be eight preproduction air vehicle systems fully integrated with all subsystems and sensors (except for two electro-optical/infrared [EO/IR] sensors); two ground segments capable of supporting the air vehicle segments; and the provision of logistical support and planning for a user-conducted two-year field demonstration of the Tier II+ system. This phase would include an irrevocable offer to supply ten air vehicle segments under Lot 1 of Phase IV for the recurring Unit Flyaway Price (UFP) of $10 million in FY 1994 dollars.

- **Phase IV** called for open-ended serial production of air vehicles 11 and subsequent and ground segment 4 and subsequent.

The apparent expectation was that at the end of the Phase III demonstration, the design would be ready for immediate serial production and operational use in Phase IV.

Not unexpectedly, Tier II+ program execution differed somewhat from the original plan. Five contractor teams rather than the planned three participated in the Phase I design competition. A budget cut just prior to the Phase II downselect resulted in the selection of only a single contractor team—led by Teledyne Ryan Aeronautical—to proceed into Phase II, rather than the originally planned two contractor teams.[1] Cost increases led the activity content of the program to be changed, and schedule slip resulted in an extended engineering flight test effort coupled with a shortened Phase III demonstration and validation effort. Nevertheless, the Global Hawk system did demonstrate sufficient military utility to transition both to the formal acquisition process and to the operational force.[2]

---

[1]Now known as the Northrop Grumman Ryan Aeronautical Center. We refer to this contractor as "Ryan" throughout the balance of this document.

[2]For details, see Robert S. Leonard and Jeffrey A. Drezner, *Innovative Development: Global Hawk and DarkStar in the HAE UAV ACTD—Program Description and Comparative Analysis*, MR-1474-AF, Santa Monica: RAND, 2001; Jeffrey A. Drezner and Robert S. Leonard, *Innovative Development: Global Hawk and DarkStar—Flight Test in the HAE UAV ACTD Program*, MR-1475-AF, Santa Monica: RAND, 2001; and Jeffrey A. Drezner and Robert S. Leonard, *Innovative Development: Global Hawk and*

Tier III– DarkStar Program Plan

The Tier III– program was a sole-source effort from its inception. The DARPA program office elected to award the Tier III– program to the Lockheed/Boeing team on the basis of prior work related to the Tier III concept. The initial Agreement between the program office and Lockheed was signed in June 1994, before DARPA and DARO had completed the process of defining the complete Tier III– program structure. As a result, the Agreement simply defined the initial phase of the program, referred to as Phase II throughout this report. The Agreement called for the design and production of two proof-of-concept flight vehicles, one radar sensor, one EO sensor, data links, and one launch control and recovery station (LCRS). No specific follow-on activities were described, but the Agreement stated the desire to rapidly and cost-effectively transition into production.

In July 1994, DARPA and DARO signed a memorandum of understanding (MOU) defining a more complete Tier III– program. Following Phase II, an additional program phase, referred to as Phase IIB, was added. Phase IIB called for the development of two to four additional air vehicle systems. As in the Global Hawk program, the apparent expectation was that the design at the end of Phase IIB would be ready for immediate serial production and operational use immediately following the ACTD. The complete original program schedule is outlined in Figure 1.2.

DarkStar program execution also differed from the original plan. The first air vehicle flew somewhat ahead of schedule but crashed during takeoff for its second flight. Flight testing resumed 26 months later. The performance of the air vehicle in flight differed from predicted characteristics, causing some concern. Costs, including contractor cost share, increased significantly. After just six flights of the second air vehicle and before Phase III began, the Air Force canceled the program.

The Global Hawk effort was aimed primarily at designing and building a system that could evolve into an operational weapon system,

DarkStar—Transitions Within and Out of the HAE UAV ACTD Program, MR-1476-AF, Santa Monica: RAND, 2001.

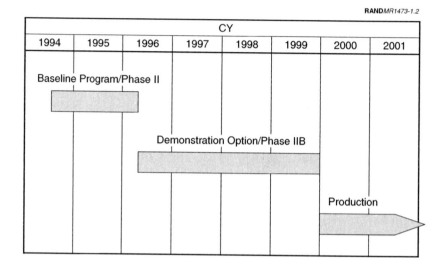

RAND*MR1473-1.2*

Figure 1.2—Tier III– Schedule, Original Complete Program

while DarkStar focused more on the task of demonstrating a technology.[3] The difference in program objectives and the corresponding differences in the anticipated abilities of the two air vehicle types were profound.

The HAE UAV program was designated an ACTD in the first year Congress authorized this development strategy. At their most aggregate level, ACTD programs are intended to provide a means for the rapid, cost-effective demonstration of new capabilities. Given a positive military utility assessment (MUA), an ACTD should accelerate the introduction of these capabilities into the military services. Most program participants believe that this objective was achieved

---

[3]At the start of DarkStar development in June 1994, the government and contractor agreed on a program that looked much like traditional technology demonstration. In fact, those very words are used in the original Agreement signed by the contractor and government.

for the Global Hawk program. Most also believed that this goal would not have been attained in the DarkStar program even if it had been allowed to complete its ACTD program.

Chapter Two

HAE UAV PROGRAM EXECUTION AND OUTCOMES

THE HAE UAV MANAGEMENT APPROACH WAS HIGHLY INNOVATIVE

The HAE UAV ACTD program adopted a radical management approach. While planning and design activities for both the Tier III– (DarkStar) and Tier II+ (eventually Global Hawk) were ongoing at the time these innovations were adopted, most were fully incorporated into the program structure as early as the May 1994 Industry Brief. The elements of this approach included the following:

- Use of Section 845 Other Transaction Authority (OTA) provided a blanket waiver for all DoD acquisition-related regulations and program management procedures, Federal Acquisition Regulation (FAR)-based processes (e.g., Cost Accounting Standards), and key legislation (e.g., the Truth-in-Negotiations Act and the Competition in Contracting Act). It also gave remarkable freedom to the government program office (DARPA at that time) to design a program structure and management approach that differed significantly from traditional approaches.

- The program's designation as an ACTD placed firm boundaries on total cost and schedule; streamlined reporting and oversight; and introduced an objective of demonstrating operational concepts and military utility.

- A UFP of $10 million each (FY 1994 dollars) for air vehicles 11–20 was the single program requirement. This price included all flight hardware: airframe, avionics, sensor(s), communications, integration and checkout, and contractor profit.

- Performance parameters were stated as goals rather than as requirements, allowing for a high degree of design flexibility.

- The use of Integrated Product Teams (IPTs) facilitated a collaborative work environment between the multiple contractors who were designing and building each system as well as between the government and the contractors who were managing each system development effort.

- The government program office intentionally remained small in order to encourage industry to manage the program efficiently.

- A high degree of contractor design authority and management responsibility encouraged efficiency as well as both technical and management innovation.

- Early user involvement was encouraged to determine military utility at the earliest possible point in the system's life cycle, well before a commitment to production was made.

Increased contractor design and management authority resulted from the program office's approach toward implementing OTA. Early user involvement is one of the core characteristics of an ACTD program. We identify these elements separately because each significantly affected the course of the program and the resulting outcomes.

At the time, no other program used this combination of initiatives in its acquisition strategy. Indeed, with the exception of IPTs, few existing programs had used any of these initiatives. As a result, there was little or no experience base on which government and industry program managers could draw to guide them in executing the HAE UAV ACTD. The management and business processes developed during the ACTD thus tended to be highly tailored to the perceived demands of the program and tended to reflect the experience and preferences of government and industry officials.

Staying within the initially estimated total program cost and schedule was given higher priority than accomplishing all the activities set out at the program's inception. This objective was also placed above achieving the technical characteristics of the system to be developed. The ACTD plan gave the program a total budget that was treated as a firm cap along with a schedule with a firm end date. Although neither of these constraints was met in a strict sense, both were only slightly exceeded. The program was thus brought to a conclusion, and a decision was made to move into the next program phase even though the ACTD had not accomplished all the tasks set out for it at its inception. Despite such changes, the program did achieve its primary objective: it demonstrated the military utility of a new capability. As a result, Global Hawk is now transitioning into follow-on development, and the first production aircraft are scheduled to be delivered in FY 2003.

ACTIVITY CONTENT CHANGED SUBSTANTIALLY FROM THE PLAN

What occurred in the HAE UAV program was in effect a continuous change of activity content throughout the ACTD in an attempt to stay within the total cost and schedule constraints defined at its inception. The Tier II+ program suffered a significant budget cut just prior to its transition into Phase II. As a result, only a single contractor was selected as opposed to the planned two. However, the management approach was not changed to compensate for the lack of competition. The program office did explicitly state that the Tier II+ and Tier III– were now in competition for future funds. Subsequently, the inherently uncertain and risky activities required to design, build, and perform basic testing of the first two aircraft in both the Global Hawk and DarkStar system development efforts ended up consuming a much larger portion of the allotted budget and calendar time than had been called for in the initial plan. To stay within the ACTD's cost and schedule constraints, testing efforts were greatly curtailed and the number of systems (Global Hawk and DarkStar air vehicles and sensor payloads) was reduced by more than half. As a result, not all operational capabilities that the system might be capable of were given sufficient opportunity for demonstration.

The activities accomplished in the HAE UAV ACTD program brought the Global Hawk system to a level of developmental maturity not equivalent to any milestone in the standard acquisition process. At the end of ACTD, Global Hawk was not a fully developed system but was well along in development. Although it was not ready for production and did not demonstrate all that was called for in the original plan, its development activities provided a strong knowledge base on the technical and performance characteristics of the system. More significantly, the Global Hawk ACTD activities accomplished the primary objective of the program: to demonstrate the military utility of a new capability. As a result of its positive MUA, Global Hawk transitioned to the formal acquisition process and to the operational force.

The early assessment for DarkStar was that the system could not be made militarily useful within known affordability constraints. By contrast, it was concluded that Global Hawk could be made operationally suitable and militarily useful given a follow-on engineering and manufacturing development (EMD) program taking less time and funding than normally required in a traditional EMD program.

COST TO THE GOVERNMENT REMAINED STABLE WHILE COSTS GREW AND ACTIVITY CONTENT CHANGED

In the course of the program, the costs of specific phases and associated activities increased significantly, but the total cost to the government remained relatively stable. The original estimated cost for the entire HAE UAV ACTD program, dated December 1994, was $912 million (in TY dollars). As of August 2000, the final estimate for all ACTD activities was $963 million (in TY dollars). This small cost difference disguises the previously discussed reductions in what was originally planned.

Calculating the cost growth of the ACTD's individual program segments—Global Hawk, DarkStar, and the common ground segment (CGS)—as well as its phases (Phases I, II, IIB, IIC, and III) is complicated. The CGS further complicates this effort because it was not part of the original plan. The preponderance of CGS costs are associated with the incorporation of DarkStar functionality into the Global Hawk ground segment and with the purchase of ground segments to service both air vehicle types.

Of the ten Global Hawk and up to ten DarkStar air vehicles in the original plan, just five Global Hawk and four DarkStar air vehicles were purchased. Fewer sensor payloads were purchased than originally planned, and some planned functions were not developed or tested (e.g., a survivability suite for Global Hawk). Only a fraction of the originally planned flight hours were flown during the much-shortened demonstration and evaluation (D&E) portion of the ACTD.[1]

Table 2.1 provides a summary analysis of the Tier II+/Global Hawk portion of the HAE UAV ACTD. The figures shown are based on a set of assumptions that allow rough comparisons of the planned and actual Tier II+ ACTD programs. The actual cost to the government of contractor work for this portion of the ACTD is slightly less than planned. However, normalizing for changes in activity content—that is, estimating the cost to complete the original plan's work scope given what we now know about costs—puts the cost at more than double the original estimate.

Table 2.2 provides similar data for the Tier III–/DarkStar portion of the HAE UAV ACTD. The actual cost for all DarkStar contractor efforts grew almost 50 percent over the original estimate. Only the Baseline Program phase can be content-normalized for comparison because the program's second phase, the Demonstration Option or Phase IIB, was not completed. The content-normalized cost for the

Table 2.1

Tier II+ ACTD Cost Analysis

Equivalent Program Phase	Actual Cost (millions of TY dollars)	Cost of Original Plan (millions of TY dollars)	Actual Cost Growth (%)	Actual Cost Normalized (millions of TY dollars)	Normalized Cost Growth (%)
Phase I	20	12	66.7	12	0
Phase II	238	230	3.5	476	100
Phase III	243	270	–10.0	623	130
Total	501	512	–2.1	1111	122

---

[1]See Leonard and Drezner, *Innovative Development: Global Hawk and DarkStar in the HAE UAV ACTD—Program Description and Comparative Analysis,* 2001, for a detailed explanation of the changes and associated costs.

Table 2.2

Tier III– Phase Cost Analysis

Equivalent Program Phase	Actual Cost (millions of TY dollars)	Cost of Original Plan (millions of TY dollars)	Actual Cost Growth (%)	Actual Cost Normalized (millions of TY dollars)	Normalized Cost Growth (%)
Baseline Program	219.8	121.5	80.9	191.9	57.9
Demonstration Option	104.3	95.5	9.2	N/A	N/A
Total	324.1	217.0	49.4	N/A	N/A

Baseline Program phase is estimated to have grown by almost 60 percent.

Overall, while the top-level budget was only slightly increased, cost increased substantially for many nonrecurring engineering development activities, necessitating reductions in the quantity of systems produced and in flight activity. Despite these changes, the program accomplished its main objective of demonstrating military utility. That assessment was positive for the Global Hawk system and (implicitly) negative for DarkStar.

OVERALL SCHEDULE GREW ONLY MODERATELY

The program's schedule outcomes were similar to its cost outcomes. Interim milestones slipped substantially, but the program's end date slipped only modestly.

The first DarkStar flew several months ahead of plan but was destroyed on takeoff on its second flight. The crash was due at least in part to elements of the acquisition strategy, particularly the use of inadequate system engineering processes. These inadequacies were enabled by the program's OTA and by the associated increase in contractor responsibility. After a 26-month hiatus, the second DarkStar air vehicle flew six more times before the program was terminated. The crash of the first DarkStar changed the tone of the HAE UAV ACTD program across all of its elements, significantly increasing risk aversion for the duration of the ACTD.

Global Hawk Phase II began as planned in April 1995. Phase III was scheduled to begin in December 1997 with the simultaneous transfer of management from DARPA to the Air Force. Problems associated with software development and system integration, along with increased risk aversion, delayed Global Hawk's first flight by 14 months. The management transition to the Air Force did not occur until October 1998. The first Phase III D&E flight was in June 1999. At that point, development and fabrication activities begun under Phase II and Phase IIB were ongoing. To stay within budget and meet unforeseen programmatic challenges, the Global Hawk program bore little resemblance to the original program plan.

In that plan, the ACTD was to be complete in December 1999 with the release of an MUA and a force-mix decision. The latter was critical to defining future program plans for both Global Hawk and DarkStar. The originally planned end date was abandoned when it became clear that slips in interim milestones had reduced the remaining time for D&E activity to the point at which sufficient information to inform the MUA could not be generated. Phase III D&E flights extended through May 2000. The original Phase III plan called for 24 months of D&E flights; actual D&E flights occurred during nine months over the 12-month period of June 1999 through May 2000.

Specifying the end date of the ACTD is dependent on how completion is defined; there are four possible dates. The final Phase III flight was July 2000.[2] If this date is used, there was a seven-month slip from the plan. The Global Hawk MUA was released in September 2000, constituting a nine-month slip from the plan. Activities put on contract during the ACTD program and considered part of the ACTD effort continued through February 2001, constituting a 14-month slip from the plan. The Milestone II decision denoting transition to the formal acquisition process was declared on March 6, 2001, making the schedule slip slightly more than 14 months. Depending on which definition is adopted, the program's total duration was between 77 months (February 1994 through July 2000) and 85 months (February 1994 through March 2001).

---

[2]For complete details, see the appendices in Drezner and Leonard, *Innovative Development: Global Hawk and DarkStar—Flight Test in the HAE UAV ACTD Program*, 2001.

UFP WAS NOT MET BUT DID HELP CONTROL COSTS AND REQUIREMENTS

The program's single requirement, the $10 million UFP (in FY 1994 dollars) that was levied on air vehicles 11–20 for both DarkStar and Global Hawk, was unattainable and ultimately abandoned. We do not view the failure to achieve the UFP as indicative of program failure. The reasons the program's sole requirement was not met are many, but we see them as falling into three categories:

- **Little or no analytical basis for the support of the original UFP.** This was the result of a deliberate philosophy of setting the price at what was believed the customer was willing to pay rather than at what actual costs would be for Tier II+ and Tier III– systems with the desired capabilities.

- **Rationalization of the UFP through extremely optimistic and essentially unrealistic assumptions.** These unfulfilled assumptions resulted in direct cost increases for components that make up the air vehicles themselves as well as in direct cost increases for running the manufacturing and engineering organizations executing the program.

- **The unwillingness of government program management to strictly enforce the cost control philosophy defined at the program's inception.** During the early portion of the program, when the system's capabilities were mostly defined, the DARPA program office was unwilling to give up any major system capability to comply with the UFP requirement.

No serious analysis underlay the UFP. To our knowledge, this number was not connected to desired Tier II+ or Tier III– capabilities in any analytical sense. Instead, we believe that the $10 million UFP was selected because it was judged to be high enough to provide a system with meaningful capability if adhered to, yet at the same time low enough that the Air Force would be willing to pay it. The originators of the program believed that the price must be set at an artificially low level or the program would be abandoned even before it began. The DoD was compelled to use this tactic because of the false notion embedded in Air Force culture that UAVs are inherently less complicated to develop and build than manned aircraft with similar capabilities.[3]

A number of specific assumptions on which the UFP was based did not come to pass. Many of these assumptions were not under the control of the DARPA or Air Force program offices. The original manufacturing schedule—which called for two aircraft in Phase II, eight aircraft in Phase III, and the ten aircraft applicable to the UFP in Phase IV—was not adhered to. Actual manufacturing activity built the intended two aircraft in Phase II but then built only three in Phase III followed by two more in the post-ACTD Phase IIC. Current planning calls for the next 12 aircraft—air vehicles 8–19—to be built at a rate of two per year over the next six fiscal years (FY 2002 through FY 2007). Air vehicles 11–20 will not be built in one lot at the production rate originally envisioned.

Further assumptions that were not met are as follows: Air vehicles 11–20 will not be built under an OTA contractual arrangement as originally assumed; the post-ACTD program will not be executed under a separate "strategic business unit"; Global Hawk will not be manufactured at an "off-site facility" to make possible so-called low-cost production; and the assumed parallel sales of key air vehicle and sensor suite components only partially materialized.

The acquisition strategy also called for the contractor to create the initial design and to control the configuration throughout the ACTD. The former occurred, but the latter was not followed as had been intended. While the DARPA program office was willing to back off on many of the system's desired capabilities, it was unwilling to remove major functionality. When it became clear that the UFP could not be attained without seriously degrading the performance of the system and when the contractor suggested doing just that, the government would not allow it.

The need for the EO/IR suite onboard the Global Hawk was the key source of disagreement between the contractor and the DARPA program office regarding the capabilities that were required to demonstrate military utility. In the 1997 time frame, Ryan believed that this

---

[3]This notion has faded in Air Force culture over the past seven years. However, it was the conventional wisdom of the time when the program began in 1994.

payload should be omitted in order to meet the UFP and that the user would find the system acceptable without this capability. The DARPA program office was steadfast in its opposition to dropping the EO/IR payload, and thus Ryan felt it had no choice but to keep the EO/IR suite regardless of its implications for the UFP requirement.

In the end, Ryan's belief was correct. As a result of the destruction of Global Hawk air vehicle 2 and Global Hawk air vehicle 3's taxi accident, no useful imagery was provided by the EO/IR payload throughout the D&E phase. Despite the system's inability to demonstrate this major capability, Global Hawk received a favorable military utility assessment.

The DARPA program office's direction to redesign Global Hawk's wing made the UFP even more unattainable. Although Ryan disputed the necessity for this redesign, the contractor ultimately succumbed to pressure from the program office and designed and produced the more robust and costly wing.

Although the UFP was not met and was ultimately abandoned, it did play two important and positive roles throughout the execution of the HAE UAV ACTD program. First, it instilled within the contractors a degree of cost consciousness that would not otherwise have materialized. Changes to system design or capabilities had UFP consequences that could be tracked. Second, the UFP provided a mechanism to prevent requirements creep. Additional requirements could not be imposed on the program without adversely affecting the UFP.

FLIGHT TESTING WAS DOMINATED BY THE NATURE OF THE SYSTEM AND THE OPERATIONAL DEMONSTRATIONS

The nature of the system in large part defined both the flight test program and the relatively slow accumulation of flight hours. The acquisition strategy—particularly the contractor's increased role in program design, management, and execution—largely defined how the flight test program was executed.

The first flight of DarkStar air vehicle 1 was on March 29, 1996. It exhibited significant anomalies that were not sufficiently resolved by the time of the second flight on April 22, 1996. The air vehicle crashed on takeoff, resulting in its total loss. DarkStar air vehicle 2 first flew 26 months later. As Figure 2.1 shows, DarkStar flight testing with air vehicle 2 lasted for a period of six months, during which only six hours of flight time were accumulated. DarkStar was terminated one day before its planned seventh flight and never entered Phase III D&E. It is not known if the shortened flight test program produced useful results either in technical characteristics relating to the air vehicle configuration or in lessons learned regarding operating procedures.

Flight testing in the Global Hawk program differed considerably from that of other air vehicle types mainly because the system is an autonomous UAV and was used in operational demonstrations. The innovative acquisition approach appears to have had little effect on the number of sorties; nor did it have an appreciable effect on the accumulation of flight hours for basic and follow-on engineering development or for air vehicle checkout flight testing.

As shown in Figure 2.2, the first flight of Global Hawk occurred on February 28, 1998. Phase II engineering flight testing included the first 21 sorties. Phase III flight testing included sorties 22–58. The

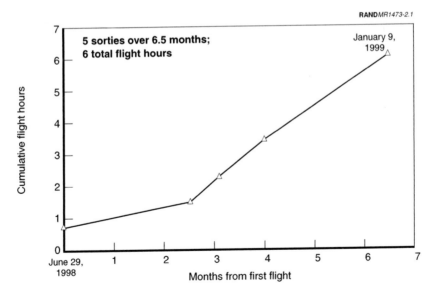

Figure 2.1—DarkStar Air Vehicle 2 ACTD Flight Test Program

Phase III flight test program included D&E sorties associated with a specific military exercise as well as some additional functional checkout and follow-on engineering flights.

No more than two air vehicles participated in the Global Hawk ACTD flight test program at any one time because air vehicle 2 was destroyed several months before air vehicle 3 became operational, and the air vehicle 3 postflight taxi accident occurred prior to the first flight of air vehicle 4. Global Hawk's 29-month ACTD flight test program reflects a clear pattern of building confidence in the system; the pace of flight testing increased significantly as the program proceeded.

The CGS did not play a role in the shortened DarkStar flight test program, as DarkStar air vehicles 1 and 2 used their own uniquely developed ground segment. The CGS was used solely in its Global Hawk–only configuration, as DarkStar air vehicles 3 and 4 never flew. With DarkStar canceled, there was no opportunity to ascertain whether the "common" aspect of the CGS functioned adequately. The lack of

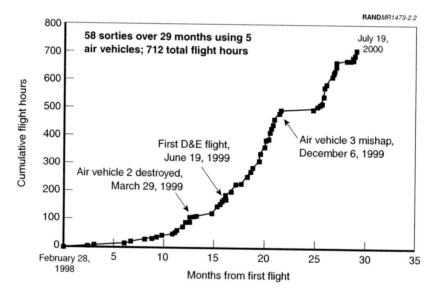

Figure 2.2—Global Hawk ACTD Flight Test Program

trained personnel and spares for Global Hawk allowed just one air vehicle in flight at any one time, and thus the CGS never demonstrated the ability to control multiple air vehicles simultaneously.

Some participants believe that neither the content of the flight test program (what was done) nor the approach used (how it was done) was greatly affected by the acquisition strategy. Evidence suggests that the dominant influence on the test program was the nature of the system. Until the HAE UAV ACTD program, very little experience had been accumulated with large autonomous UAVs. System characteristics determined the pace of the program, the profile in which flight hours were accumulated over time, and the scope of envelope expansion testing. The acquisition approach did, however, influence some key elements of the test program: the increased contractor responsibility for test program planning, direction, and execution; the early operational testing in the form of user demonstrations; and the explicit exploration of operational concepts and requirements.

PERFORMANCE GOALS WERE MOSTLY ATTAINED

The Global Hawk system came fairly close to attaining the level of performance that had been anticipated at the outset and in fact demonstrated useful capabilities that had not been anticipated. DarkStar was terminated before sufficient performance information could be generated. The CGS did not demonstrate the ability to control multiple vehicles, but it did competently control both the air vehicle and the imagery sent by it as well as the dissemination of that imagery.

The ACTD program demonstrated Global Hawk's autonomous capability of high-altitude, long-duration flight. The synthetic aperture radar sensor was capable of providing high-quality imagery, but the EO/IR sensor was not characterized sufficiently to allow an assessment of its true capability. Most of the originally specified communications and data links were demonstrated. Many Global Hawk performance parameters were close to the predicted goals, but some fell short in significant areas. In particular, a 16 percent increase in empty weight and lower-than-expected aerodynamic performance resulted in a 20 percent endurance shortfall (32 hours versus 40 hours) and a 7.7 percent shortfall in mission cruise altitude (60,000 ft versus 65,000 ft).[4]

The system also demonstrated the ability to dynamically retask both Global Hawk's sensors and, to a more limited extent, the air vehicle itself. These capabilities were not designed in and were not explicitly part of the originally conceived capabilities. However, the Air Force's 31st Test and Evaluation Squadron took advantage of its familiarity with the system, its operational perspective, and the system's inherent flexibility to create these useful capabilities. This kind of performance innovation is rarely seen during the development phases of Major Defense Acquisition programs (MDAPs).

THE INNOVATIVE ACQUISITION STRATEGY HAD A SIGNIFICANT EFFECT

We believe that the acquisition strategy used in the HAE UAV ACTD program had significant effects—both positive and negative—on program execution and outcomes. These effects could often be linked to a particular element of the strategy, but there were interactive considerations as well. Sometimes the various elements reinforced each other in their effects on the program; at other times they had a conflicting influence on the program.

- **Use of Section 845 OTA.** The use of OTA added a high degree of flexibility and tailoring to the relationship between the government and contractors. How this flexibility manifested itself and whether it led to positive or negative outcomes were in large part a function of the skills, experience, and personalities of the program managers and contracting officers. This reliance on the capabilities of individuals is a key difference between OTA and a more traditional contracting strategy. Additionally, the system engineering processes used early on in both the DarkStar and Global Hawk programs were inadequate given the complexity of the task at hand—a situation enabled by the use of

---

[4]The current Global Hawk configuration can sustain altitudes above 65,000 ft for only short periods of time, and only under specific weight and environmental conditions.

OTA and by specific wording in the Global Hawk Agreement.[5] Adequate system engineering processes were not in place until midway through Phase II in both air vehicle programs. On the plus side, OTA did eliminate costs associated with DoD- and FAR-compliant processes and reporting. It also allowed for quick and nonbureaucratic responses both to changes in the external environment and to changes prompted by internally generated learning. Combined with the use of an IPT structure, OTA facilitated a collaborative work relationship between government and contractor personnel.

- **Designation as an ACTD.** The cost and schedule boundaries associated with the ACTD designation appear constraining for a system development effort as complex as those of the Global Hawk and DarkStar. The ACTD designation resulted in significant challenges for follow-on acquisition planning and budgeting; there was a serious disconnect with the requirements and budgeting processes as the program transitioned to the formal acquisition process. This particular state of affairs can be expected in all ACTDs initiated without an approved operational requirement. Of course, the ability to begin a program without an approved operational requirements document is a key strength of the ACTD construct. Yet differing expectations regarding the entry point into the formal acquisition process created serious tension among program participants. Notably, however, the ACTD construct did provide a mechanism for demonstrating innovation—the introduction of a new capability into the operational forces—that almost certainly would not otherwise have occurred given the absence of a constituency for the HAE UAV system type within the operating services.

- **The use of a UFP as the single requirement.** The $10 million UFP in FY 1994 dollars for units 11–20 was unattainable because it was not well substantiated through analysis; its underlying assumptions were eventually invalidated; and the DARPA program office was unwilling to remove major functionality. However, the

[5]See Jeffrey A. Drezner, Geoffrey Sommer, and Robert S. Leonard, *Innovative Management in the DARPA High Altitude Endurance Unmanned Aerial Vehicle Program: Phase II Experience*, MR-1054-DARPA, Santa Monica: RAND, 1999.

UFP did promote a degree of cost consciousness—as well as a resistance to requirements creep—not normally seen in programs large enough to be considered major defense acquisitions.

- **Performance parameters stated as goals.** This allowed a degree of design flexibility that helped control costs. However, the ACTD effort forced the contractor and the government to emphasize certain system aspects while shortchanging others. Some capabilities deemed desirable later in the effort, such as faster mission planning and easier system supportability, were not emphasized during the ACTD and were therefore found lacking during the D&E effort. These attributes were consciously not made part of the ACTD program design. The flexibility of performance requirements added undesirable ambiguity to the MUA process because clear priorities were not set. This also contributed to difficulties in defining an operational requirement as the Global Hawk program transitioned to the formal acquisition process.

- **Use of IPTs.** The IPT structure generally supported the positive aspects of several other elements, including OTA and increased contractor authority. IPTs with both government and contractor personnel were used extensively in the Global Hawk program. Their use was occasionally troublesome early on in the program, but over time, as difficult personalities were removed and organizational cultures adjusted, their effectiveness improved considerably. In the DarkStar program, the contractor was not inclined to allow government individuals onto their IPTs. In addition, neither of the two DarkStar contractors participated in the other's IPTs, which almost certainly exacerbated the program's difficulties. With two equal partners, the DarkStar program was unable to make timely and effective decisions because the different cultures and competitive perspectives resulted in tension between the contractors.

- **Small program office.** Both the government (DARPA and Air Force) program offices were relatively small throughout the program. Industry program offices were also kept relatively small. The Air Force program office grew in order to accommodate planning for the transition to the formal acquisition process but remained small in relation to other high-visibility Air Force programs.

- **Contractor design authority and management responsibility.** OTA implementation in this program gave the contractors increased authority and responsibility for design, trade-off analyses and decisions, development activities, and the test program. While this appeared to have some benefits in accelerating decisionmaking and enhancing business process flexibility, both programs suffered from contractor inadequacies. In the DarkStar program, the lack of internal process discipline led to the air vehicle's problematic aerodynamic characteristics. In the Global Hawk program, inattention to system integration early in the program led to unnecessary cost and schedule increases. Where the contractor was put in charge of areas in which it lacked core competency (e.g., Ryan's responsibilities with some aspects of the flight test program), difficulties resulted.

- **Early user involvement.** This core aspect of ACTD programs designated the Joint Forces Command (JFCOM) as the user in the HAE UAV ACTD. An operational demonstration test phase supporting an MUA was an essential element of the program structure. Early user involvement showed clear benefits in providing focus and emphasis in this ACTD and in guiding the tailoring of the development effort to follow. However, unified commands such as JFCOM do not have the resources or the expertise to be fully interactive in the development process. The operational user—in this case the Air Force's Air Combat Command (ACC)—does have the necessary resources and expertise, but the ACC did not become involved until post-ACTD planning had commenced. These circumstances resulted in considerable tension between the unified commands (e.g., JFCOM) and operational user communities.

Our overall assessment is that the acquisition strategy used for the development of the Global Hawk and DarkStar systems during the HAE UAV ACTD contains innovations with strong potential for improving acquisition processes for other kinds of systems. We believe that the program's innovative management was largely responsible for the various successes of the program: It was highly flexible and responsive to changes in circumstances; was easily modified to incorporate those changes into the program; and provided a focus on demonstrating a new capability that allowed for the prioritization of competing objectives and interests. Ultimately, the strategy resulted in the introduction into the operational force of a new capability of demonstrated usefulness—an outcome that we believe to be superior to what would have occurred under a traditional approach.

Given the experience of the HAE UAV ACTD program, we believe that some modifications to elements of the overall strategy, as well as refinements in the way such a strategy is implemented, are required for broader application.

Chapter Three

COMPARISON TO OTHER PROGRAMS

HAE UAV efforts do not involve simply building a glorified model airplane or drone, as some who view UAVs as "low tech" compared to manned aircraft might imagine. On the contrary, the Global Hawk and DarkStar programs are in many respects at least as complex and challenging as typical manned aircraft developments. Each of the two HAE UAV programs involved the development of multiple system segments: an air vehicle, sensor payloads, and ground segments (one unique to each platform plus the CGS). Most major defense programs develop only one of these basic elements: the air vehicle, sensors or some other payload, or a ground station.

For purposes of comparison, the Global Hawk and DarkStar development efforts were considered roughly equivalent to those of contemporary manned military aircraft programs that created all-new air vehicle designs. For the most part, it was not possible to compare the costs of the two HAE UAV development efforts to those of recent or historical UAV programs, as we found no unclassified or declassified UAV program with both sufficient available data and sufficient complexity and technological challenge to be appropriate for cost comparison to the Global Hawk and DarkStar HAE UAVs. In our schedule comparison, we do include other UAV programs along with appropriate manned programs. In this area, we again believe that the latter are the more appropriate comparative systems.

THE SYSTEM CONCEPT AND ACTD CONSTRUCT DICTATED A UNIQUE FLIGHT TEST PROGRAM

The uniqueness of flight testing an autonomous HAE UAV complicates comparisons. The development of an autonomous UAV requires an enhanced understanding of the system earlier in testing than what is required for a manned or remotely piloted vehicle system. Manned aircraft have the benefit of the pilot's ability to identify problems through the direct feel of the air vehicle, and his or her reactions can mitigate problems during flight testing. A remotely piloted vehicle has a human in the loop if not onboard, thus offering the advantage of real-time human input. HAE UAV flight testing benefited neither from real-time human input nor from real-time human feedback from the air vehicle.

Flight test comparisons are also complicated by differences in the area and composition of flight envelopes (speed, altitude, maneuverability) for different aircraft types. The high-altitude and long-range mission profile of the HAE UAVs is specifically and narrowly defined, resulting in an air vehicle flight envelope with a very small area.[1] Other aircraft types have missions that rely on great aerodynamic flexibility, dictating the use of larger-area flight envelopes that must be explored during flight test.

Figure 3.1 presents a rough comparison of aircraft flight test programs. Global Hawk's flight test experience is compared with the flight test experience of several fighter aircraft during their advanced technology demonstration and EMD development phases. We observe a very different pattern of flight hour accumulation over time in the Global Hawk ACTD. Fighter aircraft tend to accumulate hours much faster as multiple aircraft fly multiple short sorties each month. Global Hawk had only two aircraft in flight-ready status at any one time and generally flew no more than three long-duration sorties per month.

---

[1]Data on the U-2 (the system most similar to the HAE UAVs) flight test program were unavailable.

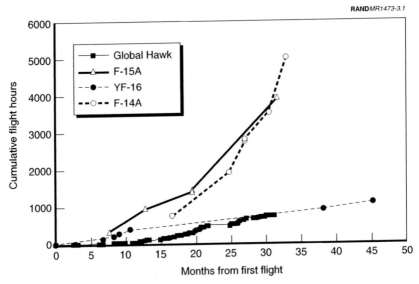

RAND*MR1473-3.1*

SOURCE: Global Hawk data are from quick-look and flash reports (as of October 2, 2000). Data on the F-14A, F-15A, and YF-16/F-16 were collected by an industry source and validated with other information collected by RAND. The F-14A data include the period from December 1970 through September 1973. The F-15A data include the period from July 1972 through March 1975. The YF-16/F-16 data include both the LWF program and some full-scale development from January 1974 through October 1977.

Figure 3.1—Comparison of Aircraft Flight Test Programs

As shown in Table 3.1, a comparison of the Global Hawk ACTD to prototype programs provides a different perspective.[2] The AX program involving two pairs of attack prototype aircraft included a very short, seven-month flight test program. The first five months were conducted by the contractors, and in the next two a formal competitive fly-off using Air Force pilots was undertaken. In the first five months, the YA-9 accumulated 162 hours using two aircraft and an additional 145.5 hours (in 123 sorties) in the two-month competitive evaluation. The competing YA-10 accumulated 190 hours in the first

---

[2]Information on the AX and LWF programs are taken from Giles K. Smith, A. A. Barbour, Thomas L. McNaugher, Michael D. Rich, and William L. Stanley, *The Use of Prototypes in Weapon System Development*, R-2345-AF, Santa Monica: RAND, 1981.

Table 3.1

Prototype Program Flight Test Comparison

Program	Testing Type	Duration (months)	Flight Hours	Sorties
Global Hawk	Engineering	6	20.5	5
YA-9	Engineering	5	162.0	N/A
YA-9	Demonstration	2	145.5	123
YA-10	Engineering	5	190.0	N/A
YA-10	Demonstration	2	138.5	87
YF-16	Program total	11	450.0	320
YF-17	Program total	7	350.0	230

five months with its two aircraft and 138.5 hours (87 sorties) in the two-month fly-off. The YA-10s were then used for 16 months during full-scale development (FSD) for follow-on development test and evaluation (DT&E) and initial operational test and evaluation (IOT&E) prior to the low-rate production decision. In contrast, Global Hawk accumulated 20.5 flight hours in five sorties in the first six months of its flight test.

The Lightweight Fighter (LWF) program of the early 1970s was a highly streamlined effort involving two pairs of prototype fighter aircraft. In 11 months in 1974, the YF-16 accumulated approximately 450 hours in 320 sorties. The YF-17 flew 230 sorties and accumulated approximately 350 hours in a seven-month period that same year. Clearly the pace of the LWF program flight test was much faster than that of Global Hawk. This is due in part to Global Hawk's system type—a large autonomous UAV. Global Hawk requires much more preparation for each mission, specifically in the area of mission planning.

Both competitive prototype flight test programs described above had informal MUAs through the participation of operationally oriented test pilots. The introduction of JFCOM in the Global Hawk ACTD as the architect of the D&E phase, with a focus on the formal demonstration of UAV as a joint warfighting asset, differed from the "ACC-equivalent" user approach involved in these comparative programs from the 1970s.

In general, traditional fighter aircraft EMD flight test programs fly more sorties per month and accumulate more flight hours than did

Global Hawk during its ACTD. The main explanation for this difference is the vastly larger flight envelope for fighter systems and the
need to satisfy myriad requirements, as is traditionally the approach
in an MDAP-compliant EMD effort.

MORE WAS ACCOMPLISHED AT A SOMEWHAT LESSER COST

Development activity through the end of the ACTD and into follow-
on development efforts was examined in two segments: the early
portion embodied in the ACTD's Phase II, which involved the design
and build of the first two air vehicles, and the remainder of the ACTD
along with proposed follow-on development activity. The former
applies to both the Global Hawk and DarkStar programs; the latter
applies only to the Global Hawk program. Analysis of DarkStar beyond its Phase II is not possible, as its Phase IIB—the building of
follow-on aircraft—was not completed.

Figure 3.2 compares the costs of the two HAE UAV ACTD Phase II
efforts to those of selected manned military aircraft demonstra-
tion/validation (dem/val), prototype and technology demonstration
programs of the past 30 years.[3] The systems in Figure 3.2 represent
experimental fighter, attack, cargo, and reconnaissance aircraft.[4]
Roughly similar basic activities are found in all of these programs regardless of whether the air vehicle was intended to evolve into an operational system.

For the six programs showing two segments to their cost bar, the
lower segment represents the program's cost excluding flight test activity, which is accounted for in the upper segment. The OTA contractual arrangements in the Global Hawk and DarkStar programs
made it difficult to break out flight test costs from those of other activities during Phase II; thus, a single-segment bar is shown for total
cost of the phase. Tacit Blue data were available only in aggregate;

---

[3]The Advanced Tactical Fighter dem/val program is not shown. The cost of each of
the YF-22 and YF-23 programs was over $2 billion in TY dollars (when prime and
subcontractor investments are included).

[4]The Have Blue technology demonstration program informed the development of the
F-117, which is essentially an attack aircraft.

hence the single bar. The costs of flight testing in the YF-16 and YF-17 prototype programs did not come out of these programs' budgets and were not separately accounted for. As a result, these flight test costs are not available in the historical record and are not included in Figure 3.2.[5]

Total expenditures in the programs shown in Figure 3.2, escalated to FY 2001 dollars, range from a low of less than $200 million to a high of more than $500 million. The average costs for the design-and-build portion and for the flight test portion of the historical programs for which such costs can be determined are $275 million and $92 million, respectively.[6] By our calculations, Global Hawk Phase II cost $238 million paid to contractors plus allocated government costs of roughly $40 million. DarkStar cost $220 million in payments to contractors plus allocated government costs of some $37 million.[7] Converted to FY 2001 dollars, the Global Hawk and DarkStar totals come to $295 million and $273 million, respectively.

Each program shown in Figure 3.2 accomplished a unique set of activities. Our opinion is that the best comparative programs for DarkStar are Have Blue and Tacit Blue and for Global Hawk the LWF prototypes. Given the activity content of Phase II and the pending operational demonstration for the HAE UAVs at the conclusion of that phase, we believe that the developmental maturity of the HAE UAVs at the end of Phase II of the ACTD was higher than that of any comparative program listed in Figure 3.2.

---

[5]See Lieutenant Colonel Morris R. Betry, UASFR, "The History of Technology Viability, Technology Demonstrator, and Operational Concept Prototype Program Costs," July 1994.

[6]The average for the design-and-build portion is calculated from these costs in the X-31, X-29, YC-15, YC-14, YF-17, YF-16, Have Blue, and XFV-12 programs. The average for the flight test portion is calculated from this cost in the X-31, X-29, YC-15, YC-14, Have Blue, and XFV-12 programs.

[7]See Leonard and Drezner, *Innovative Development: Global Hawk and DarkStar in the HAE UAV ACTD—Program Description and Comparative Analysis*, 2001, for a full explanation of costs. As of January 1999, total government costs during the ACTD were estimated at $138 million. Global Hawk Phase II accounted for 29 percent of all payments to contractors in the ACTD; thus, that percentage of government costs is allocated to the effort. For DarkStar, the figure is 27 percent.

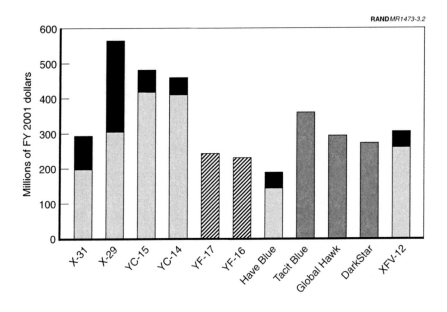

Figure 3.2—Program Comparison: Equivalents to Phase II of the HAE UAV ACTD

The final cost of DarkStar Phase II was roughly what one would have expected given the costs and accomplishments of comparable historical programs such as Have Blue and Tacit Blue. We found that the Global Hawk Phase II effort compared favorably to the LWF prototypes once each was adjusted for known definitional differences in their estimates. This is a favorable outcome given that conventional wisdom views the LWF prototype program as one of the most successful prototype programs in Air Force history.

A different and perhaps more important comparative metric lies in estimating the value of the overall acquisition strategy in reducing future development costs. We define the Global Hawk equivalent EMD program as ACTD Phases IIB and III; Phases IIC and the pre-EMD, which bridge the ACTD to the formal EMD; and the planned Spirals 1 and 2 in EMD. The cumulative costs of these six phases are compared to EMD expenditures in the F-16A/B, F/A-18A/B, and F-117A programs. The projected cost of the Global Hawk–equivalent EMD is about $1.1 billion in FY 2001 dollars. This is only slightly more than half the inflation-adjusted EMD costs in the least expensive of the comparison programs, the F-16, and slightly more than one-third more than the inflation-adjusted expense of the most appropriate EMD comparison program, the F-117. F/A-18A/B EMD costs were the highest of all the other programs.

A large band of uncertainty surrounds the future of Global Hawk development expenditures. However, after realistic upper and lower bounds for Global Hawk's equivalent EMD figure have been defined, its cost will be considerably less than what one might expect given historical programs of roughly similar technological challenge and system complexity.

MILITARY UTILITY WAS DEMONSTRATED MORE QUICKLY

The Global Hawk ACTD concluded with a system more developmentally mature than what is typical at the conclusion of a prototype, an advanced technology demonstrator, or a dem/val program. At the same time, the Global Hawk ACTD did not attain the system maturity typically seen at the conclusion of a traditional EMD program.

Figure 3.3 compares in some detail the overall schedules for selected programs.[8] The time to the first flight of the prototype (☆) and the time to the first flight of an EMD test aircraft (O) are indicated.[9] The length of time from program start to EMD initiation for Global Hawk and DarkStar[10] was roughly comparable to that in the F-117, the F-16, and Compass Cope.[11] Global Hawk's time to first flight (prototype) was somewhat longer than that for the F-16, F-117, and Compass Cope. DarkStar's timing is more comparable. Global

---

[8]The length of time from EMD initiation to first production aircraft delivery is estimated in all but two cases.

[9]Note that the Medium-Altitude Equivalent UAV Predator ACTD did not build what we consider to be prototype aircraft. The Gnat-750—from which Predator was derived—acted in that capacity.

[10]We use our "equivalent EMD" analytical approach here. The initiation of Phase IIB with the building of the third and subsequent aircraft in each HAE UAV ACTD program signifies EMD initiation.

[11]Compass Cope was a competitive dem/val program of the early 1970s that produced two flying examples of two different HAE UAV air vehicles. The aircraft did not fly autonomously, and the program included no sensor suite or ground station development efforts.

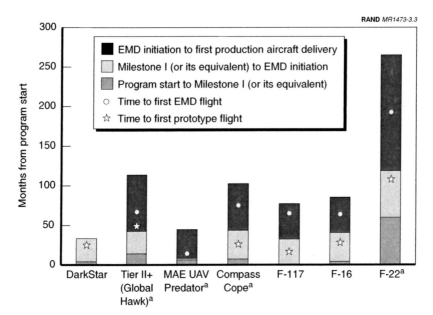

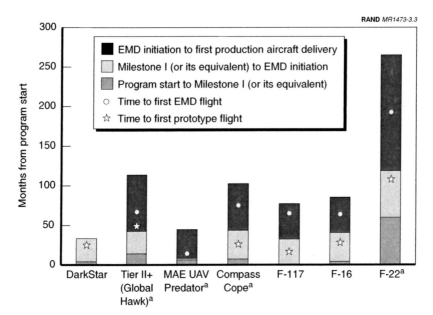

^aFirst operational delivery estimated.

Figure 3.3—Schedule Comparison

Hawk's time to first EMD flight is similar to that of the other three programs (this date is estimated in the Compass Cope program).

The most striking difference in the Global Hawk program is the short time span between the first flight of the first prototype aircraft and that of air vehicle 3, which is considered the first EMD-equivalent aircraft. This resulted directly from the ACTD acquisition strategy and allowed what was learned in the building of the first two aircraft to be fully leveraged in the manufacture of the third and subsequent aircraft. Similar production continuity is expected between the building of the final EMD aircraft, air vehicle 7, and the first production aircraft, air vehicle 8.

The F-22 schedule as shown in Figure 3.3 is extremely long compared to any of the other programs shown. The schedule is shown as a reference for just how long the development cycle can be when a traditional development approach is used for the most sophisticated of weapon system programs.

We conclude that the ACTD acquisition strategy in the HAE UAV ACTD program did not result in an accelerated first flight for either the prototype or the EMD aircraft. However, the core of an ACTD flight test program is its emphasis on quickly demonstrating military utility. Global Hawk did so very early in its flight test program. By the end of the Phase III D&E, Global Hawk had demonstrated its capability for long-distance deployment. Except for the Predator UAV, which was also initially developed under the ACTD acquisition strategy and similarly benefited from OTA designation, no other aircraft demonstrated both utility and deployment capability so early in its test program. By these measures, which are at the core of the ACTD concept, the Global Hawk program was successful.

Chapter Four

LESSONS AND RECOMMENDATIONS

ACTD PROGRAMS QUICKLY MATURE TECHNOLOGY AND DEMONSTRATE NEW OPERATIONAL CAPABILITIES

We believe that the ACTD program achieved a level of developmental and system maturity beyond what a traditional program would accomplish given similar time and funding. Yet while much was accomplished, at the end of the HAE UAV ACTD Global Hawk was not a fully developed system and was not ready for production and sustained operational deployment.

The Air Force program office believes that had the program's overriding priority lay in making the system production ready by the completion of the ACTD, this goal could have been accomplished. However, the overall objective of the acquisition strategy was to demonstrate an enhancement to the warfighter's operational capabilities in a way that was deemed by the users to be worth the cost. Given this overriding objective, the path to success in the ACTD required the use of limited program funding to prove the system's military utility via the execution of the D&E program.

There is a bias in ACTD guidance toward transitioning directly into low-rate production at the conclusion of an ACTD. This bias was reflected in the HAE UAV's original plan and subsequent post-ACTD planning efforts. We believe that this bias is neither suitable nor appropriate for a large, complex system. The constrained budget and tight schedule of an ACTD program are not conducive to addressing the complete development needs of a major defense system. As a result, many programs with characteristics that place them in the

realm of an MDAP—i.e., those with expected development and procurement costs large enough to be classified as an MDAP—may be
too complex and involve too much risk to be constrained by the program structure an ACTD acquisition strategy demands. An ACTD's
focus is rightly oriented toward demonstrating military utility, not
toward the operationalization of a system.

ACTD-developed systems with characteristics that place them in the
realm of an MDAP should be biased toward entry into the traditional
process at EMD. The EMD should be tailored to take advantage of
what was learned in the ACTD. Planning for the transition from
ACTD to the tailored EMD should begin midway through the ACTD,
thus providing enough time to ensure that the benefits of the ACTD
experience become the foundation for post-ACTD development activities.

THE INNOVATIVE MANAGEMENT APPROACH WAS RESPONSIVE TO CHANGE

The HAE UAV ACTD program illustrates an important characteristic
of program management that could benefit a wider array of programs. The management approach employed in the HAE UAV ACTD
program was highly responsive to changing circumstances, including
funding issues, technical challenges, cost increases, oversight conditions, political support, and business environment. Changes could
be made to the program structure, the timing and length of phases,
and performance characteristics that reflected both changes in the
external environment and lessons from the ongoing program. There
was also sufficient flexibility in the strategy to adjust development
and test activity content in response to such stimuli while retaining a
focus on the primary objective of demonstrating the utility of a new
capability. That flexibility was facilitated by the use of OTA and IPTs,
which led to a collaborative working relationship, as well as by the
ACTD construct and early user participation, which maintained and
enforced the program's prioritization of objectives.

GOVERNMENT MUST EXERT STRONG OVERSIGHT AND HAVE KEEN INSIGHT

The government program office must establish and maintain a strong oversight function while managing a program under OTA. The increased level of contractor design authority and management responsibility should not imply decreased government oversight responsibility; in fact, it demands more insight into the contractor's activities and progress. Because industry and government are motivated differently, only the government can ensure that its interests remain paramount. Even with a strong top-level oversight function, the government-contractor relationship can and must be collaborative at the working level using an IPT structure. This combination of management characteristics—government insight, strong oversight, and a collaborative work relationship—developed over time in the HAE UAV ACTD program.

Contractor responsibility should be limited to those areas constituting that contractor's core competency. Different contractor organizations have different strengths and weaknesses. The unstructured nature of OTA implementation can lead to the elimination of formal processes, particulary with regard to system engineering, that are essential to the success of a program. Care must be taken that the OTA contractual instrument is written to include equivalent processes.[1]

Government program management must both be aware of and compensate for those weaknesses on the part of its contractor or contractor team. There are some areas—specifically flight testing beyond basic engineering flight test—in which virtually no contractor has a core competency. For instance, Ryan did very well in test execution but fared poorly in test planning. In these areas, the acquisition strategy should be altered to ensure that appropriate government organizations are in charge.

The government must also retain the authority to modify program goals as information is created. While increased contractor design

---

[1]More recent programs using OTA have addressed this issue and have avoided many of the problems experienced in the early portions of the HAE UAV ACTD program. Giles K. Smith, Jeffrey A. Drezner, and Irving Lachow address this in "Assessing the Use of 'Other Transaction' Authority for Prototype Projects," DB-375-OSD, Santa Monica: RAND, forthcoming.

responsibility and management authority can contribute to innovation and streamlined program execution, the government must be able to enforce the execution of a program that addresses the government's priorities and risk preferences.

PROGRAM OBJECTIVES SHOULD BE MODIFIED AND BALANCED AS THE PROGRAM EVOLVES

Overall cost and schedule were essentially fixed in the HAE UAV ACTD program, with system performance stated as the parameter with a degree of flexibility. The only system requirement, the UFP, appears to have been designed to promote the program's survival up to the point at which the Air Force found the HAE UAV systems useful and therefore became less concerned with its cost. This strategy was successful in the Global Hawk program, as the Air Force abandoned the UFP shortly after D&E exercises proved the system's potential. The $10 million UFP instilled a cost consciousness both in the program office and at the contractor.

We believe that cost, schedule, performance, and unit price should be stated as goals to be traded off within identified bounds. This approach allows for more intelligent trade-offs and leads to more cost-effective system design solutions. When one or more of these four parameters are fixed, the program becomes irrationally constrained. Flexibility of this sort already exists, but its use requires enhancement and encouragement in the acquisition, user, and requirements communities. All of these communities need to be more accepting of innovative concepts and more accepting of the results of trade-offs among cost, schedule, and performance objectives.

OPERATIONAL USERS MUST BE INVOLVED EARLY

Users must include both operators and product users—i.e., those responsible for putting the system into operation and those requiring the results of the system's successful operation. In most cases, these users are different communities or organizations with different perspectives, cultures, and priorities. Attention must be paid to the valid requirements and desires of both users. It is the input and participation of operational users that allows trade-offs to be focused on key performance parameters that significantly affect mission capability.

Operational users must be incorporated into program decisions and processes early in the system's development, as it is these users who have the resources and knowledge to actively participate in the development process. To ensure the mutual support of both users and the success of the system under development, a mechanism should be put in place from the program's onset to quickly resolve user conflicts that are not easily resolved at the working level. We acknowledge, however, that involving operational users early in a development program is extremely challenging and that there are few examples of success. An operational user's desire to maximize operational suitability and supportability may conflict with the objective of maturing new technology and demonstrating a core capability. The challenge for program management is to achieve the appropriate balance given the objectives of the effort.

UP-FRONT PLANNING AND PREPARATION ARE CRITICAL TO SUCCESS

Sufficient time should be spent very early in the program on developing realistic cost, schedule, performance, and unit price goals as well as on creating a well-designed statement of overall program objectives to guide managers throughout the development activity. The characterization and assessment of risks, determination of nonrecurring engineering tasks, and planning for future transitions must also be conducted more thoroughly up front. These items should be revisited periodically during the course of the development effort.

Whether between phases and associated stages of development and technical maturity or of management responsibility and approach, transitions can be anticipated and provisions made for their realization. Early planning, inclusion of all relevant stakeholders, top-level support, and event-driven milestones can contribute to smoother transitions.[2]

Organizations with a current or future interest in the system under development should have significant input into early planning. Key elements of the Air Force—particularly the operational users (ACC)— did not buy into the Global Hawk program. Had they been involved up front, a smoother program execution and transition into the force structure might have resulted. Early management plans should make it clear that the designated lead agency for post-ACTD activities should plan on funding the operationalization of the system. Mid-program reviews, potentially held immediately after flight test begins, should yield considerable insight into the nature and utility of the system and the kinds of development activities required in the future. Earlier consideration of these issues should resolve most of the requirements and budgeting process mismatches that occur at the conclusion of an ACTD.

A process should be put in place to ensure that the expectations of the various organizations involved in the program are managed and are realistic. In particular, expectations regarding the possible entry point of the system into the acquisition process should be thoroughly vetted. The acquisition and user communities should recognize that an important result of this type of approach might be the transfer of knowledge—new operational concepts, ideas, and technologies. These communities should also recognize that not all systems developed elsewhere are inherently deficient.

CONSISTENT SENIOR-LEVEL INVOLVEMENT AND SUPPORT IS CRITICAL TO SUCCESS

The consistent involvement and support of senior decisionmakers from key stakeholders is a critical factor in an ACTD's successful program execution. These endorsements facilitated the transition of the system into the force structure as well as the management approach for conducting further development, test, and production.

---

[2]For a complete discussion in the HAE UAV context, see Drezner and Leonard, *Innovative Development: Global Hawk and DarkStar—Transitions Within and Out of the HAE UAV ACTD Program*, 2001.

Stakeholders must include the leaders of the developing agency, the service and OSD acquisition organizations, the testing community, the designated unified command sponsor, and the operational user. Even with consistent senior-level support, resistance to new ideas can be expected throughout the organization. Senior-level support enables such resistance to be directly addressed and eventually overcome.

BIBLIOGRAPHY

"ACTD Transition Guidelines: Executive Summary," available at www.acq.osd.mil/actd/.

"Air Force Lab Pushes UAVs for AWACS, JSTARS, RIVET Joint Missions," *Inside the Air Force*, July 21, 2000, pp. 15–17.

"Air Force to Appeal Senate Cut to Global Hawk Advance Procurement," *Inside the Air Force*, June 2, 2000, pp. 3–4.

"Air Staff to Brief Peters on Accelerating Global Hawk, Retiring U-2," *Inside the Air Force*, July 21, 2000, pp. 11–12.

"Australia Eyes UAV," *Aviation Week & Space Technology*, March 1, 1999, p. 35.

"End of UAV Demos Leaves ASC Wondering How to Budget for the Vehicles," *Inside the Air Force*, December 11, 1998, p. 7.

"Faulty Mission Preparation Cited in December Global Hawk Accident," *Inside the Air Force*, Vol. 11, No. 17, April 28, 2000, pp. 12–13.

"Global Hawk Incident Prompted Changes: Military Alters Communications Management Rules After UAV Crash," *Inside the Air Force*, January 28, 2000, p. 3.

"Global Hawk 2 Flight Sets Stage for Airborne Sensor Tests," *Aviation Week & Space Technology*, November 30, 1998, p. 32.

"Global Hawk UAV Crash Linked to Test Error," *Defense News*, January 10, 2000.

"Improper Mission Computer Input Said to Be Key Factor in UAV Crash," *Inside the Air Force*, March 17, 2000, p. 14.

"Introduction to ACTDs," available at www.acq.osd.mil/actd.

"Let's Make a Deal," *Aviation Week & Space Technology*, August 30, 1999, p. 21.

"Out, DarkStar," *Aviation Week & Space Technology*, February 1, 1999, p. 27.

"Pentagon Argues Global Hawk Cut Will Delay Fielding by One Year," *Inside the Air Force*, July 28, 2000, pp. 15–17.

"Poor Communications Management Cited in Global Hawk UAV Crash," *Inside the Air Force*, January 7, 2000, pp. 9–10.

"RQ-4A Global Hawk Unmanned Aerial Vehicle Accident," *Inside the Air Force*, April 28, 2000, pp. 13–15.

"Senate Authorizers Want to Slow Global Hawk UAV Procurement Plans," *Inside the Air Force*, May 19, 2000, pp. 7–8, 13, and 23–27.

"Senators Push for Global Hawk to Explore Airborne Surveillance Role," *Inside the Air Force*, May 12, 2000, pp. 5–6.

"Tenter, De Leon to Determine Future Airborne Recon Mix, Funding Needs," *Inside the Air Force*, August 4, 2000, p. 3.

Advanced Systems and Concepts, *ACTD Master Plan*, CD-ROM, September 2000.

Asker, James R., "Let's Make a Deal," *Washington Outlook*, August 30, 1999, p. 24.

Atkinson, David, "Global Hawk Crash Delays Demos," *Defense Daily*, April 1999.

Barraco Klement, Mary Ann, "Agile Support Project—Global Hawk Program: Rapid Supply, Responsive Logistics Support for Next-Generation UAVs," *PM*, January–February 1999, pp. 66–70.

Birkler, John, Giles Smith, Glenn A. Kent and Robert V. Johnson, *An Acquisition Strategy, Process, and Organization for Innovative Systems*, Santa Monica: RAND, MR-1098-OSD, 2000.

Congressional Budget Office, "The Department of Defense's Advanced Concept Technology Demonstrations," Washington, D. C., September 1998.

de France, Linda, "UAVs Hold Key to Future Conflicts, Kosovo Air Commander Says," *Aerospace Daily*, November 15, 2000.

Dornheim, Michael A., "Destruct System Eyed in Global Hawk Crash," *Aviation Week & Space Technology*, April 5, 1999, p. 61.

Drezner, Jeffrey A., and Robert S. Leonard, *Innovative Development: Global Hawk and DarkStar—Flight Test in the HAE UAV ACTD Program*, MR-1475-AF, Santa Monica: RAND, 2001.

Drezner Jeffrey A., and Robert S. Leonard, *Innovative Development: Global Hawk and DarkStar—Transitions Within and Out of the HAE UAV ACTD Program*, MR-1476-AF, Santa Monica: RAND, 2001.

Drezner, Jeffrey A., Geoffrey Sommer, and Robert S. Leonard, *Innovative Management in the DARPA High Altitude Endurance Unmanned Aerial Vehicle Program: Phase II Experience*, MR-1054-DARPA, Santa Monica: RAND, 1999.

Eash, Joseph, "ACTDs Link Speed, Technology for U.S. Forces," *Defense News*, September 20, 1999.

Eash, Joseph, "Two Major UAVs Show Defense Program's Success," *Aviation Week & Space Technology*, August 7, 2000, p. 74.

Fulghum, David A., "Long-Term Stealth Project Gets the Ax," *Aviation Week & Space Technology*, May 24, 1999, pp. 77–78.

Fulghum, David A., "Low-Cost Mini-Radar Developed for UAVs," *Aviation Week & Space Technology*, October 9, 2000, p. 101.

Fulghum, David A., "Recce Funding Increase Pits U-2 Against Global Hawk," *Aviation Week & Space Technology*, September 27, 1999, p. 37.

Fulghum, David A., "Will New Elusive Craft Rise from DarkStar?" *Aviation Week & Space Technology*, February 22, 1999, pp. 27–28.

Fulghum, David A., and Robert Wall, "Global Hawk Gains Military Endorsement," *Aviation Week & Space Technology*, September 18, 2000, p. 34.

Fulghum, David A., and Robert Wall, "Global Hawk Gains Military Endorsement," *Aviation Week & Space Technology*, September 18, 2000, pp. 34–36.

Fulghum, David A., and Robert Wall, "Global Hawk Snares Big Break," *Aviation Week & Space Technology*, October 23, 2000, p. 55.

Fulghum, David A., and Robert Wall, "Long-Hidden Research Spawns Black UCAV," *Aviation Week & Space Technology*, September 25, 2000, pp. 28–30.

Fulghum, David A., and Robert Wall, "New Missions, Designs Eyed for Global Hawk," *Aviation Week & Space Technology*, November 20, 2000, p. 57.

Harmon, Bruce R., Lisa M., Ward, and Paul R. Palmer, *Assessing Acquisition Schedules for Tactical Aircraft*, Alexandria, VA: Institute for Defense Analyses, IDA Paper P-2105, 1989.

Henderson, Breck W., "Boeing Condor Raises UAV Performance Levels," *Aviation Week & Space Technology*, April 23, 1990.

Hundley, Richard O., "DARPA: Technology Transitions: Problems and Opportunities," internal document, Santa Monica: RAND, June 1999.

Johnson, Robert V., and Birkler, John, *Three Programs and Ten Criteria: Evaluating and Improving Acquisition Program Management and Oversight Processes Within the Department of Defense*, MR-758-OSD, Santa Monica: RAND, 1996.

Johnson, Robert V., and Michael R. Thirtle, "Management of Class III Advanced Concept Technology Demonstration (ACTD) Programs: An Early and Preliminary View," internal document, Santa Monica: RAND, July 1997.

Leonard, Robert S., and Jeffrey A. Drezner, *Innovative Development: Global Hawk and DarkStar in the HAE UAV ACTD—Program*

Description and Comparative Analysis, MR-1474-AF, Santa Monica: RAND, 2001.

Leonard, Robert S., Jeffrey A. Drezner, and Geoffrey Sommer, *The Arsenal Ship Acquisition Process Experience: Contrasting and Common Impressions from the Contractor Teams and Joint Program Office,* MR-1030-DARPA, Santa Monica: RAND, 1999.

Mann, Paul, "Joint Ops Make Gains, but 'Jointness' Lags," *Aviation Week & Space Technology,* April 10, 2000, p. 27.

McMichael, William H., "Global Hawk Teams with U.S. Carrier Group," *Defense News,* June 19, 2000, p. 24.

Moteff, John D., "IB1022: Defense Research: DOD's Research, Development, Test, and Evaluation Program," *CRS Issue Brief for Congress,* August 13, 1999.

Mulholland, David, "Pentagon Cancels DarkStar UAV, Pursues Global Hawk," *Defense News,* 1999.

Pae, Peter, "Military Is Sold on Unmanned Spy Plane" *Los Angeles Times,* January 8, 2001.

Proctor, Paul, "Sensor Deprivation," *Aviation Week & Space Technology,* January 31, 2000, p. 17.

Pugliese, David, "Canada Delays UAV Acquisition," *Defense News,* June 14, 1999, p. 18.

Ricks, Thomas E., and Anne Marie Squeo, "The Price of Power: Why the Pentagon Is Often Slow to Pursue Promising New Weapons," *Wall Street Journal,* October 12, 1999, p. 1.

Scott, William B., "F-22 Flight Tests Paced by Aircraft Availability," *Aviation Week & Technology,* October 16, 2000, pp. 53–54.

Smith, Giles K., "The Use of Flight Test Results in Support of High-Rate Production Go-Ahead for the B-2 Bomber," internal docment, Santa Monica: RAND, 1991.

Smith, Giles K., "Use of Flight Test Results in Support of F-22 Production Decision," internal document, Santa Monica: RAND, 1994.

Smith, Giles K., A. A. Barbour, Thomas L. McNaugher, Michael D. Rich, and William L. Stanley, *The Use of Prototypes in Weapon System Development*, R-2345-AF, Santa Monica: RAND, 1981.

Smith, Giles K., Hyman L. Shulman, and Robert S. Leonard, *Application of F-117 Acquisition Strategy to Other Programs in the New Aquisition Environment*, MR-749-AF, Santa Monica: RAND, 1996.

Sommer, Geoffrey, Giles K. Smith, John L. Birkler, and James R. Chiesa, T*he Global Hawk Unmanned Aerial Vehicle Acquisition Process: A Summary of Phase I Experience*, MR-809-DARPA, Santa Monica: RAND, 1997.

Thirtle, Michael R., "Origination of the High Altitude Endurance (HAE) UAV ACTD," internal document, Santa Monica: RAND, 1998.

Thirtle, Michael R., Robert V. Johnson, and John L. Birkler, *The Predator ACTD: A Case Study for Transition Planning to the Formal Acquisition Process*, MR-899-OSD, Santa Monica: RAND, 1997.

U.S. General Accounting Office, "Best Practices: Better Management of Technology Development Can Improve Weapon System Outcomes," GAO/NSIAD-99-162, July 1999.

U.S. General Accounting Office, "Defense Acquisition: Advanced Concept Technology Demonstration Program Can Be Improved," GAO/NSIAD-99-4, October 1998.

U.S. General Accounting Office, "Unmanned Aerial Vehicles: DOD's Demonstration Approach Has Improved Project Outcomes," GAO/NSIAD-99-3, August 1999.

U.S. General Accounting Office, "Unmanned Aerial Vehicles: Progress of the Global Hawk Advanced Concept Technology Demonstration," GAO/NSIAD-00-78, April 2000.

U.S. General Accounting Office, "Unmanned Aerial Vehicles: Progress Toward Meeting High Altitude Endurance Aircraft Price Goals," GAO/NSIAD-99-29, December 1998.

Wall, Robert, "U.S. Surveillance Aircraft to Get Budget Boost," *Aviation Week & Space Technology*, January 31, 2000, pp. 32–33.

Wall, Robert, "USAF Maps Out Future of Global Hawk UAV," *Aviation Week & Space Technology*, July 12, 1999, p. 53.

Zaloga, Steven J., "Conflicts Underscore UAV Value, Vulnerability," *Aviation Week & Space Technology*, January 17, 2000, pp. 103–112.

OTHER PROGRAM DOCUMENTATION

Air Force Operational Test and Evaluation Center, Detachment 1, *Global Hawk System Advanced Concept Technology Demonstration: Quick Look Report for Roving Sands '99*, 1999.

Basic Systems for the High Altitude Endurance Unmanned Aerial Vehicle System (HAE UAV): ACAT Level I, Initial Requirements Document CAF 353-92-II, 1999.

Defense Advanced Research Projects Agency, *Advanced Concept and Technology Demonstration (ACTD) Management Plan (MP) for the Medium Altitude Endurance (MAE) Unmanned Aerial Vehicle (UAV)*, Arlington, VA, 1994.

Defense Advanced Research Projects Agency, *HAE UAV ACTD Management Plan*, Arlington, VA, December 15, 1994.

Defense Advanced Research Projects Agency, *High Altitude Endurance Unmanned Aerial Vehicle: Advanced Concept Technology Demonstration: Management Plan*, Arlington, VA, 1994.

Defense Advanced Research Projects Agency, *High Altitude Endurance Unmanned Aerial Vehicle Program (HAE UAV): Advanced Concept Technology Demonstration: Management Plan*, Arlington, VA, 1997.

Defense Advanced Research Projects Agency, *System Specification for the Global Hawk High Altitude Endurance (HAE) Unmanned Air Vehicle*, Arlington, VA, 367-0000-003E, 1997.

Demonstration & Evaluation Integrated Process Team Operations Plan, 1997.

Director of Systems Acquisition for Under Secretary of Defense (Acquisition & Technology) and Principal Deputy Under Secretary of Defense (Acquisition & Technology), Executive Summary regarding the Global Hawk DAE Review Memorandum, 1999.

Global Hawk Program Office, ASC/RAV, *Global Hawk Program Monthly Acquisition Reports*, September–December 2000.

Global Hawk Program Office, ASC/RAV, *Global Hawk Program Quads*, August–November 2000.

Global Hawk Program Office, ASC/RAV, *Global Hawk Program Single Acquisition Management Plan*, November 1, 2000.

Global Hawk Program Office, ASC/RAV, *Global Hawk System Program Overview*, December 14, 2000.

Global Hawk Program Office, ASC/RAV, *Global Hawk Update to Dr. Gansler, SAF/AQIJ*, 1999.

Heber, Charles E., Jr., *Air Combat Command's Role in the HAE UAV ACTD*, HAE UAV Program Office, 1998.

High Altitude (HAE) Unmanned Aerial Vehicle (UAV) Advanced Concept Technology Demonstration (ACTD), PMD 2404 (1)/PE# 35205F, 1999.

High Altitude Endurance Unmanned Aerial Vehicle Program Office— Aeronautical Systems Center, *Joint High Altitude Endurance Unmanned Aerial Vehicle Program (HAE UAV) Single Acquisition Management Plan*, Wright-Patterson Air Force Base, OH, 1999.

Hooper, Major Guy, handouts from presentation entitled "Global Hawk ACTD Lessons Learned," 2000.

McPherson, Colonel Craig, Global Hawk Program Director, ASC/RAV, presentation entitled "Global Hawk System: Early Strategy and Issues Session," 2000.

McPherson, Colonel Craig, Global Hawk Program Director, presentation for congressional staffers entitled "Global Hawk System," 2000.

McPherson, Craig, *RQ-4A Global Hawk: High Altitude Endurance Unmanned Aerial Reconnaissance System: Command, Control, Communications, Computers and Intelligence Support Plan,* Wright-Patterson Air Force Base, OH: Global Hawk Program Office—Aeronautical Systems Center, 2000.

Memorandum for ASC/RA from HQ AFMC/DO, Wright-Patterson Air Force Base, OH, regarding Responsible Test Organization (RTO) designation, Global Hawk System Test Program, 2000.

Memorandum for ASC/CC from SAF/AQ regarding Dark Star termination, 1999.

Memorandum for Record from Office of the Under Secretary of Defense regarding High Altitude Endurance (HAE) UAV Program Review, 1999.

Memorandum for the Secretary of the Air Force (ATTN: Acquisition Executive) from the Under Secretary of Defense regarding Global Hawk Decision Memorandum, 1999.

Memorandum for Under Secretary of Defense (Acquisition & Technology), Principal Deputy Under Secretary of Defense (Acquisition & Technology) regarding the Read-Ahead Global Hawk DAE meeting, July 7, 1999.

Sullivan, Kevin J., *High Altitude Endurance Unmanned Aerial Vehicle (HAE UAV) Monthly Acquisition Report,* 1999.

Teledyne Ryan Aeronautical, *Air Worthiness Five (5) Through Air Worthiness Seven (7) Test Missions: Detailed Test Plan for the Tier II Plus High Altitude Endurance Unmanned Aerial Vehicle System,* San Diego, CA, Report No. TRA-367-5000-157, 1998.

Teledyne Ryan Aeronautical, *Master Test Plan for the Tier II Plus High Altitude Endurance (HAE) Unmanned Air Vehicle,* San Diego, CA, Report No. TRA-367-5000-67-R-001, 1995.

Teledyne Ryan Aeronautical, *Master Test Plan for the Tier II Plus High Altitude Endurance (HAE) Unmanned Aerial Vehicle*, San Diego, CA, Report No. TRA-367-5000-67-R-001A, November 17, 1995.

Teledyne Ryan Aeronautical, *Payload Test Missions*, San Diego, CA, Report No. TRA-367-5000-202, 1998.

U.S. Atlantic Command, *HAE UAV ACTD Integrated Assessment Plan*, 1998.

U.S. Atlantic Command, *HAE UAV ACTD Joint Concept of Operations*, 1996.

U.S. Atlantic Command, *HAE UAV Joint Employment Concept of Operations*, 1998.

U.S. Joint Forces Command, *Global Hawk System Advanced Concept Technology Demonstration: Military Utility Assessment*, April 1995 to June 2000.